Charlie goes for a walk,

introducing algebra

with oranges

Sarah Reeves

Sarah's Books Pty Ltd, PO Box 1904 Armidale NSW 2350

First Edition, 2024
First published, 2024

ISBN: 978-1-7635799-4-1
Independently published.

Printed in Sydney, NSW, Australia, if purchased in Australia.
eBook also available.

A catalogue record for this book is available from the National Library of Australia.

This book was created, written and illustrated in Armidale NSW, which is Anaiwan country.

Links to the Australian Mathematics Curricula

ACARA Mathematics Curriculum link:
Foundation - AC9MFA01 - recognise, copy and continue repeating patterns represented in different ways.
Year 1 - AC9M1A02 - recognise, continue and create repeating patterns with numbers, symbols, shapes and objects, identifying the repeating unit and recognising the importance of repetition in solving problems.

NSW Mathematics K-10 Curriculum link:
Early stage 1 - MAE-FG-02 - forms equal groups by sharing and counting collections of objects.

Victorian Mathematics F - 6 Curriculum link:
Foundation - VC2MFA01 - follow a short sequence of instructions; recognise, copy, continue and create repeating patterns represented in different ways.
Level 1 - VC2M1A02 - recognise, continue and create repeating patterns with numbers, symbols, shapes and objects, identifying the repeating unit and recognising the importance of repetition in solving problems.

Links to curricula:
The Australian Curriculum, Assessment and Reporting Authority (ACARA) Mathematics F-10 Version 9.0
The NSW Education Standards Authority (NESA) Mathematics K-10 Curriculum (February 2024)
The Victorian Curriculum and Assessment Authority Mathematics Foundation to Level 6 V 2.0

This book is dedicated to David.

Thank you for always being there.

Charlie goes for a walk.

Charlie finds 1 orange.

Charlie finds another

2 oranges.

🍊 + 🍊 + 🍊 = 3 🍊

Charlie finds 1 more orange.

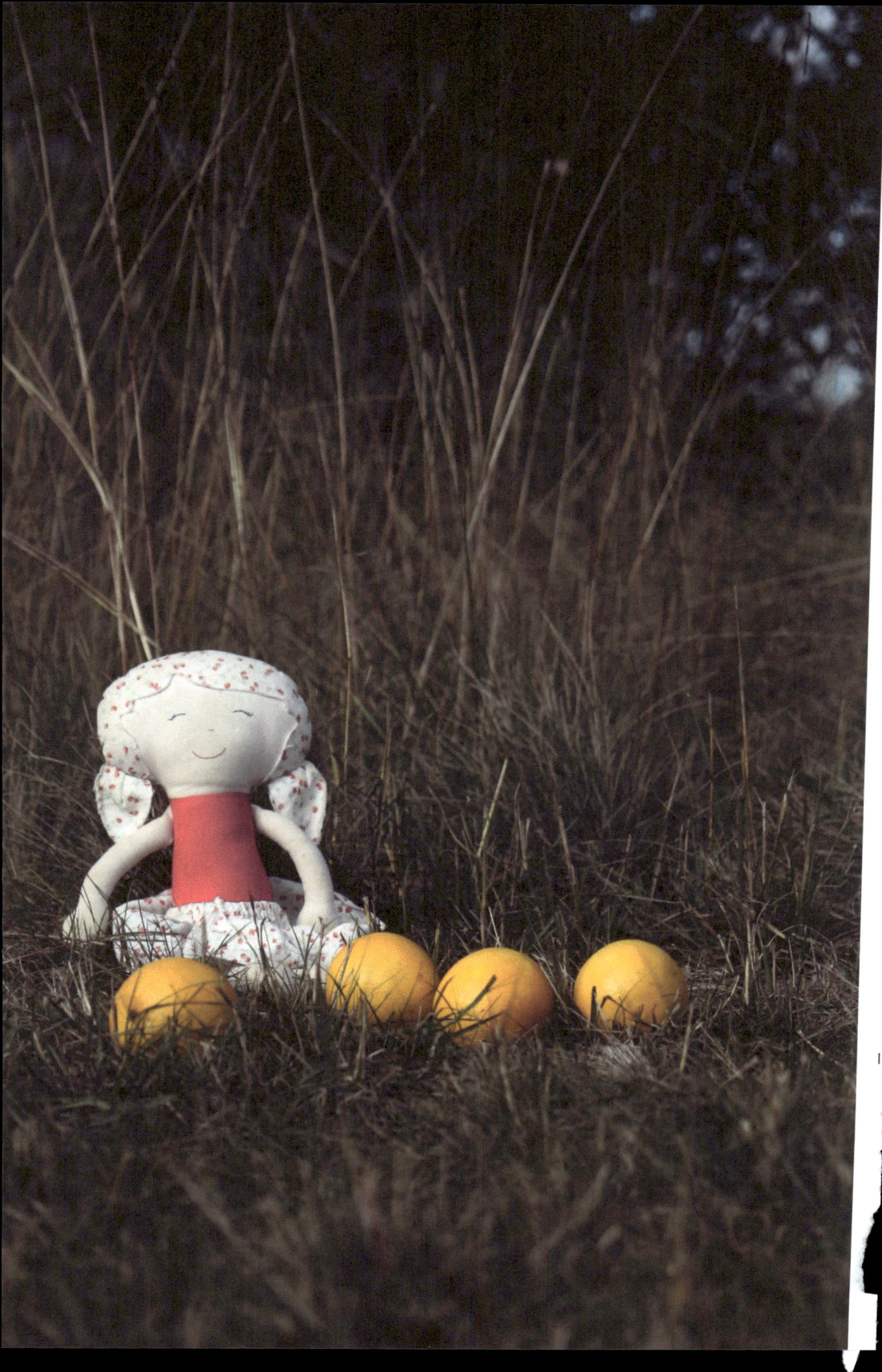

Charlie now has 4 oranges.

🍊 + 3 🍊 = 4 🍊

Charlie finds ANOTHER

orange.

🟠 + 4 🟠 = 5 🟠

Charlie decides to eat

1 orange.

5 🍊 - 🍊 = 4 🍊

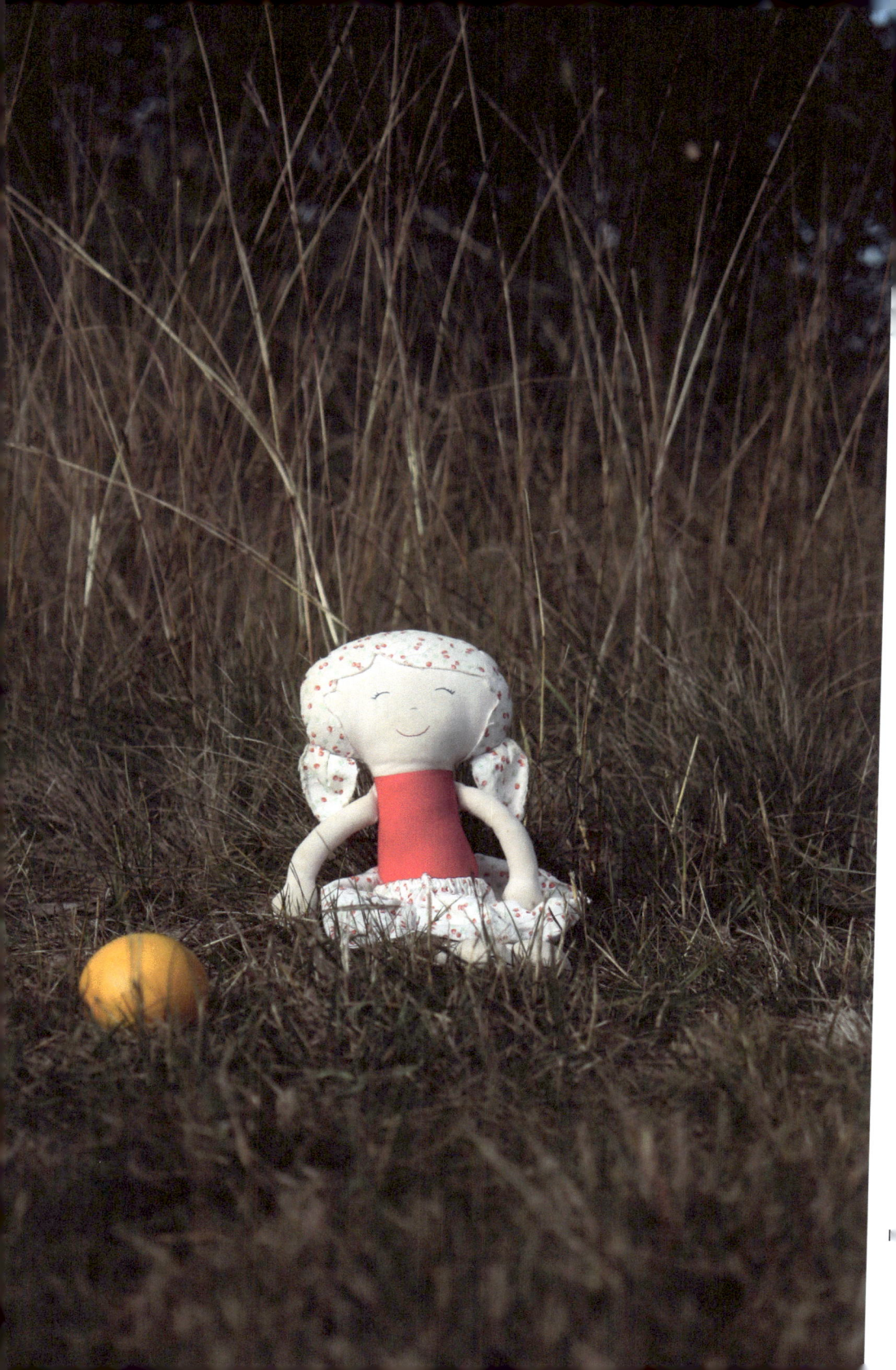

It was so delicious,

Charlie decides to eat

2 more.

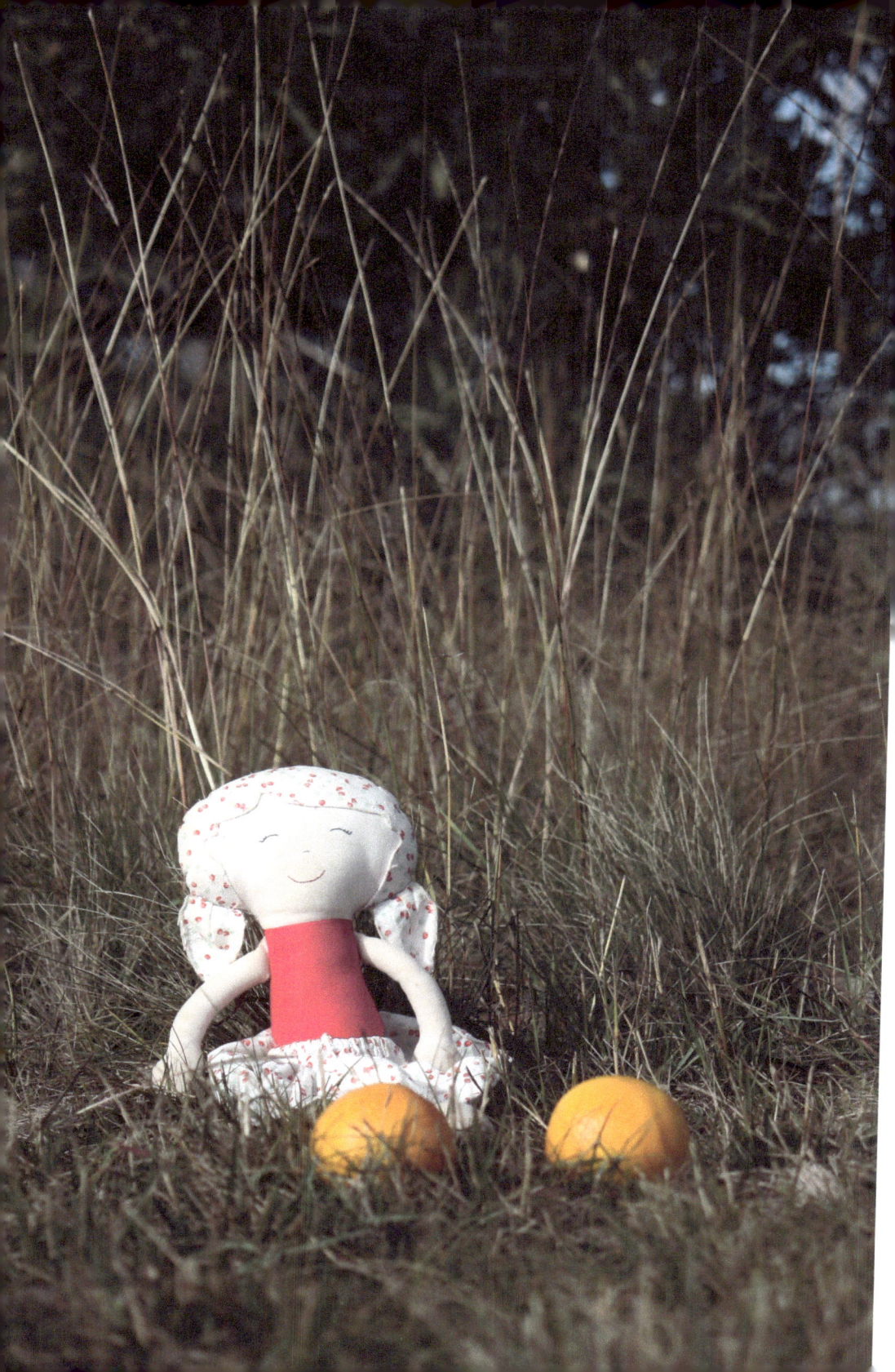

Charlie is left with 2 oranges

to take home.

4 🍊 - 2 🍊 = 2 🍊

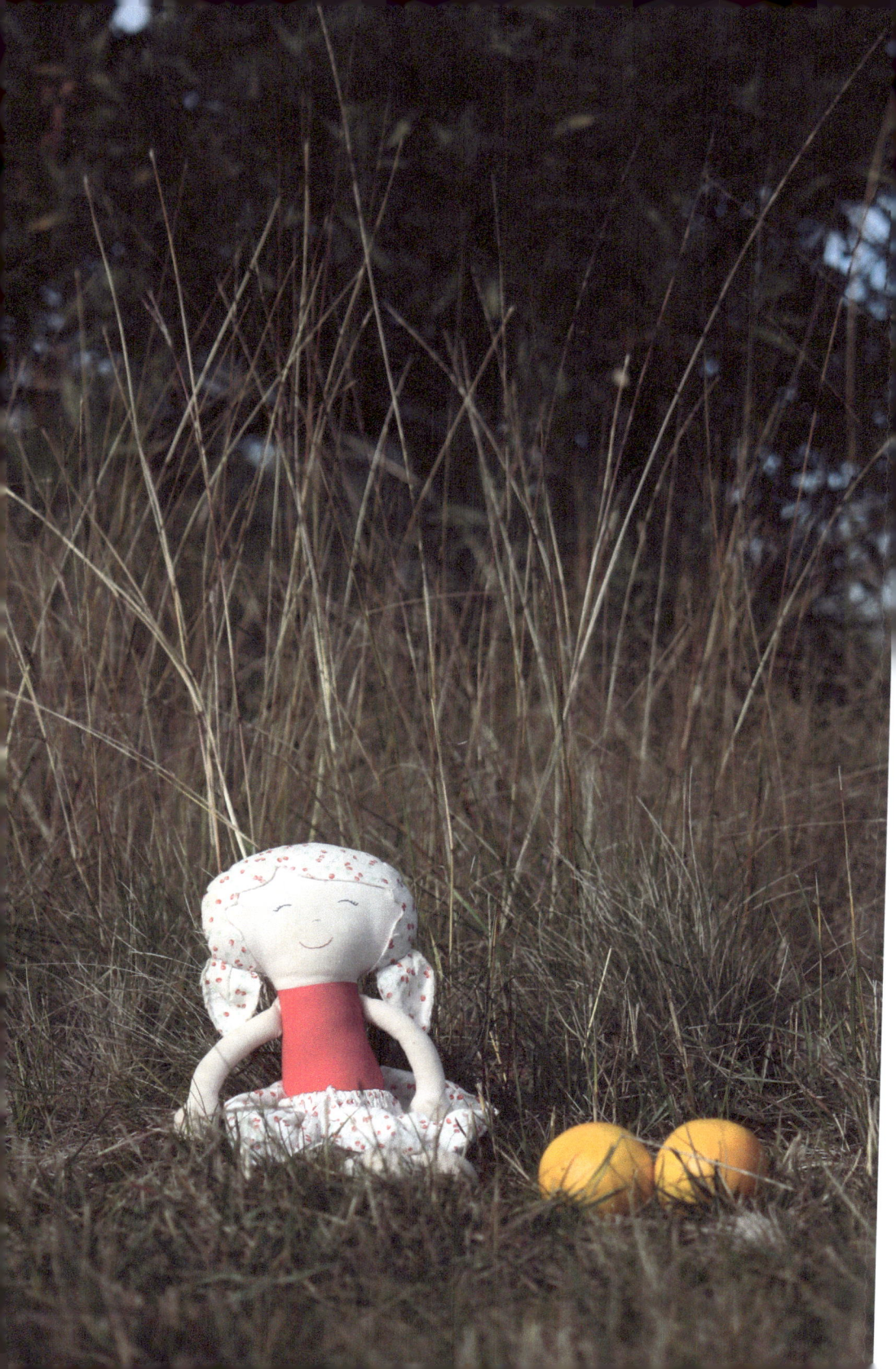

Charlie has used the symbol

to represent an orange.

This makes adding and

subtracting the number of

oranges much simpler.

The End.

Other books available in the Charlie series:

Charlie counts to five, on a picnic
ACARA Mathematics Curriculum link AC9MFN02
NSW Mathematics K-10 Curriculum link MAE-RWN-001 &
MAE-RWN-002
Victorian Mathematics F - 6 Curriculum link VC2MFN02

Charlie explores whole number 7
ACARA Mathematics Curriculum link AC9MFN06
NSW Mathematics K-10 Curriculum link MAE-RWN-01
Victorian Mathematics F - 6 Curriculum link VC2MFN06

Charlie loves to share, exploring odd and even numbers
ACARA Mathematics Curriculum link AC9MFN06
NSW Mathematics K-10 Curriculum link MAE-FG-02 & MA1-FG01
Victorian Mathematics F - 6 Curriculum link VC2MFN06 & VC2M3N01

Charlie looks into greater than, less than and equal to
ACARA Mathematics Curriculum link AC9MFN06
NSW Mathematics K-10 Curriculum link MAE-RWN-02
Victorian Mathematics F - 6 Curriculum link VC2MFN01

Charlie finds a pattern, in the sandpit
ACARA Mathematics Curriculum link AC9MFA01
NSW Mathematics K-10 Curriculum link MAE-FG-01
Victorian Mathematics F - 6 Curriculum link VC2MFA01

Other books available in the Charlie series:

Charlie loves to cook, exploring measurement
ACARA Mathematics Curriculum link AC9MFM01
NSW Mathematics K-10 Curriculum link MAE-CSQ01
Victorian Mathematics F - 6 Curriculum link VC2MFM01

Charlie explores fractions, parts of a whole
ACARA Mathematics Curriculum link AC9M2M02
NSW Mathematics K-10 Curriculum link MAE-GM-03
Victorian Mathematics F - 6 Curriculum link VC2M2M02

Charlie finds Geometry, in nature
ACARA Mathematics Curriculum link AC9MFSP01
NSW Mathematics K-10 Curriculum link MAE-2DS-01 & MA1-2DS-01
Victorian Mathematics F - 6 Curriculum link VC2MFSP01

Links to curricula:
The Australian Curriculum, Assessment and Reporting Authority (ACARA)
Mathematics F-10 Version 9.0
The NSW Education Standards Authority (NESA) Mathematics K-10
Curriculum (February 2024)
The Victorian Curriculum and Assessment Authority Mathematics Foundation
to Level 6 V 2.0

www.ingramcontent.com/pod-product-compliance
Lightning Source LLC
Chambersburg PA
CBHW042120060426

42449CB00030B/39